The House Book

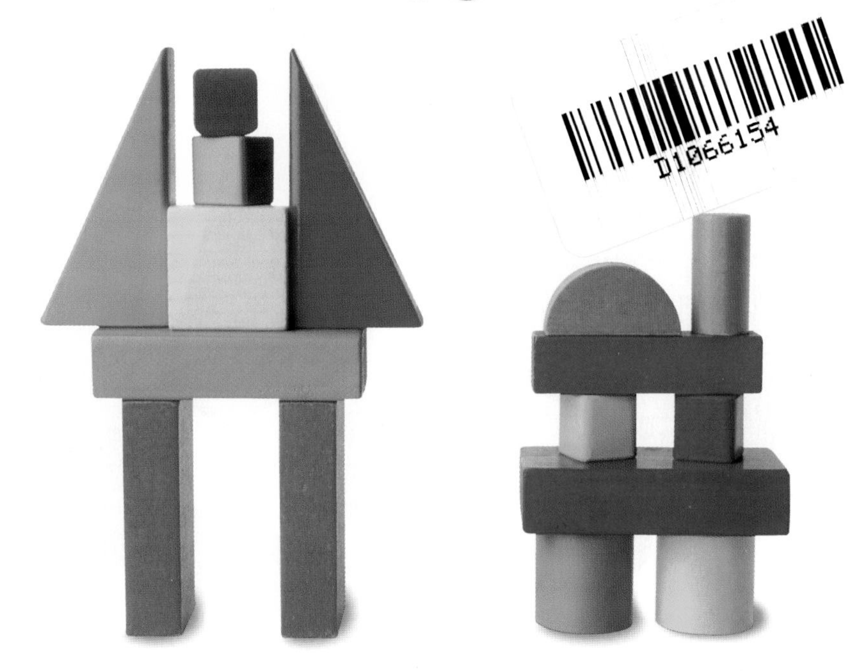

written by Shirley Frederick

Orlando Boston Dallas Chicago San Diego

www.harcourtschool.com

my blocks

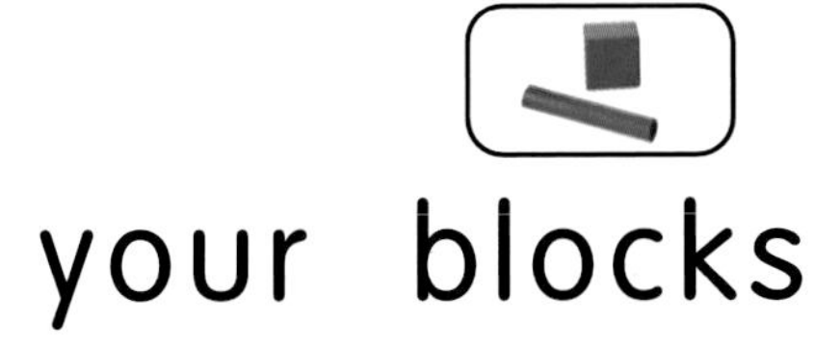

your blocks

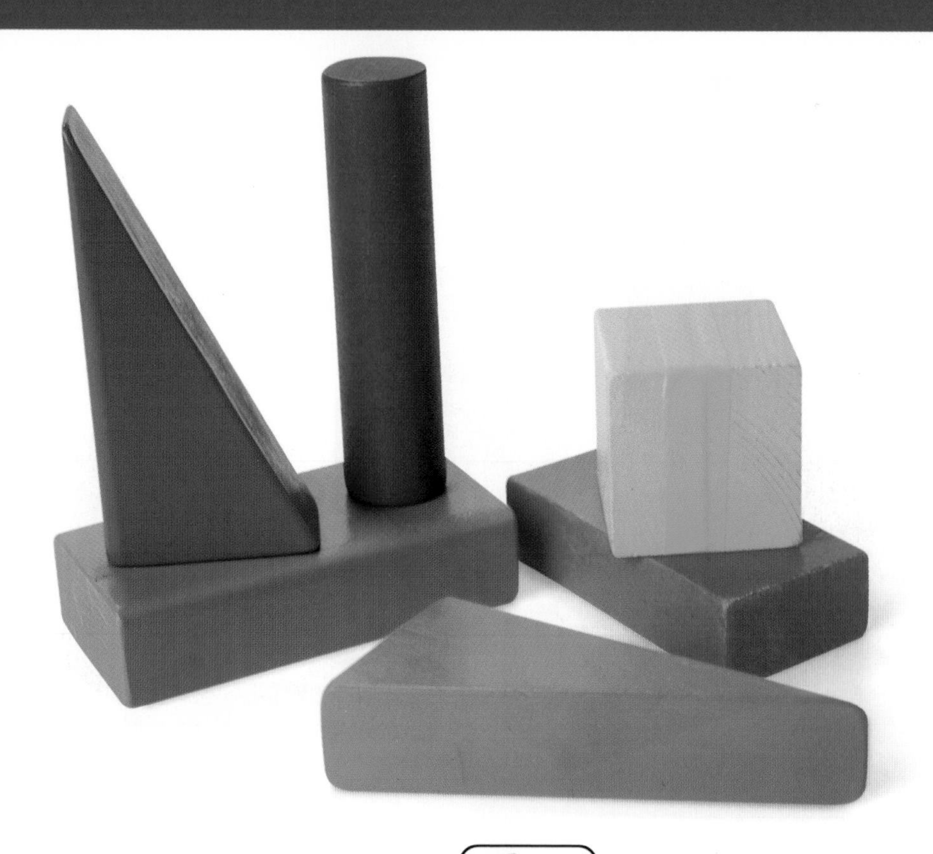

my blocks

your blocks

my blocks

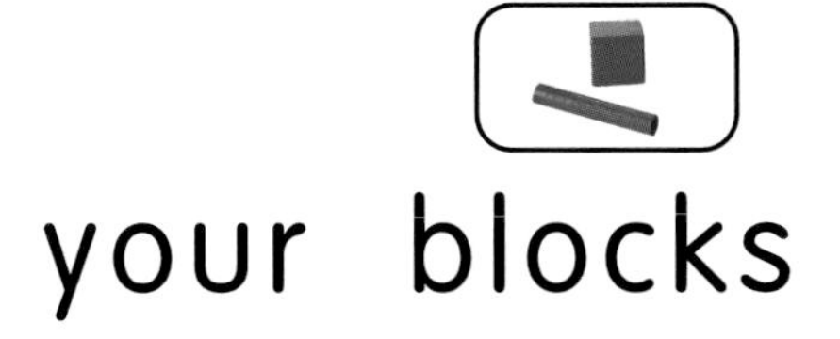

your blocks

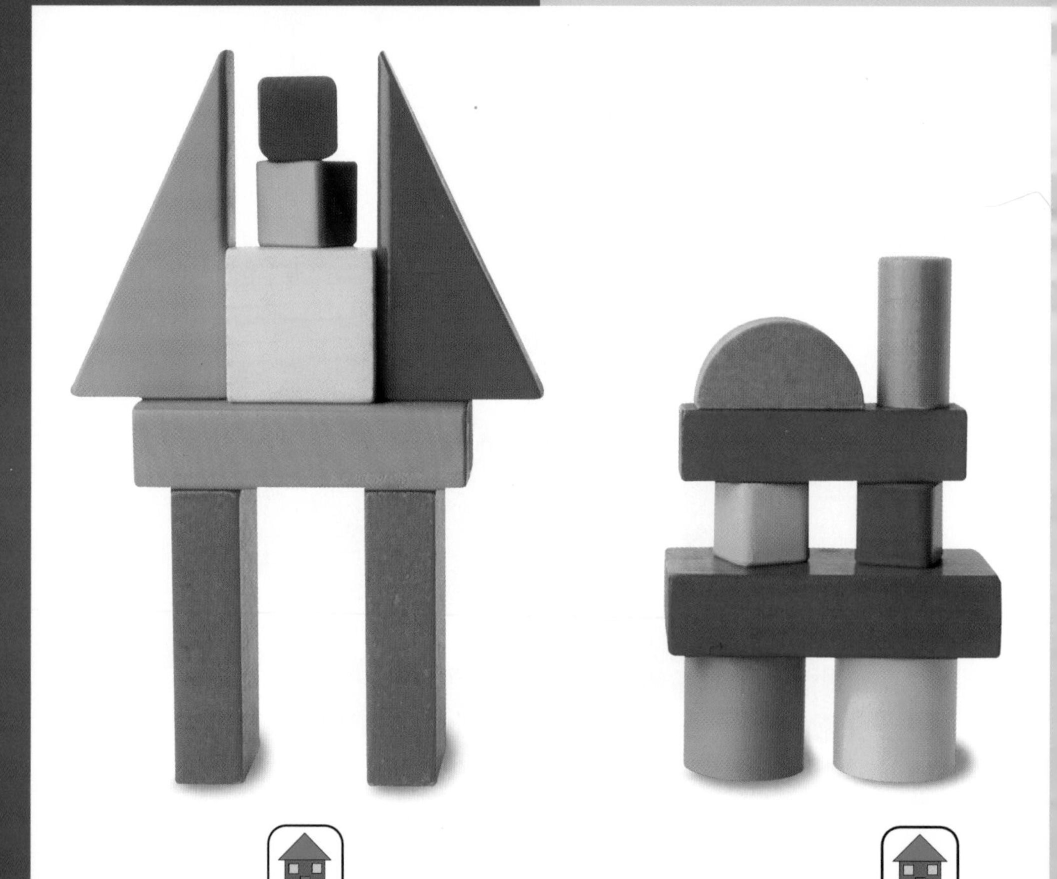

my house and your house

Teacher/Family Member

Invite your child to build with blocks. Then ask him or her to identify attributes of the blocks he or she used, for example, long, curved, square.

Sight Word: *your*

Word Count: 17

All photos by Sheri O'Neal / Harcourt.

Printed in the United States of America

Grade K Book 5 ISBN 0-15-314846-2

Ordering Options: 0-15-316197-3 Package of 5
0-15-316225-2 Grade K Package, Books 1-6

6 7 8 9 10 179 2001 00

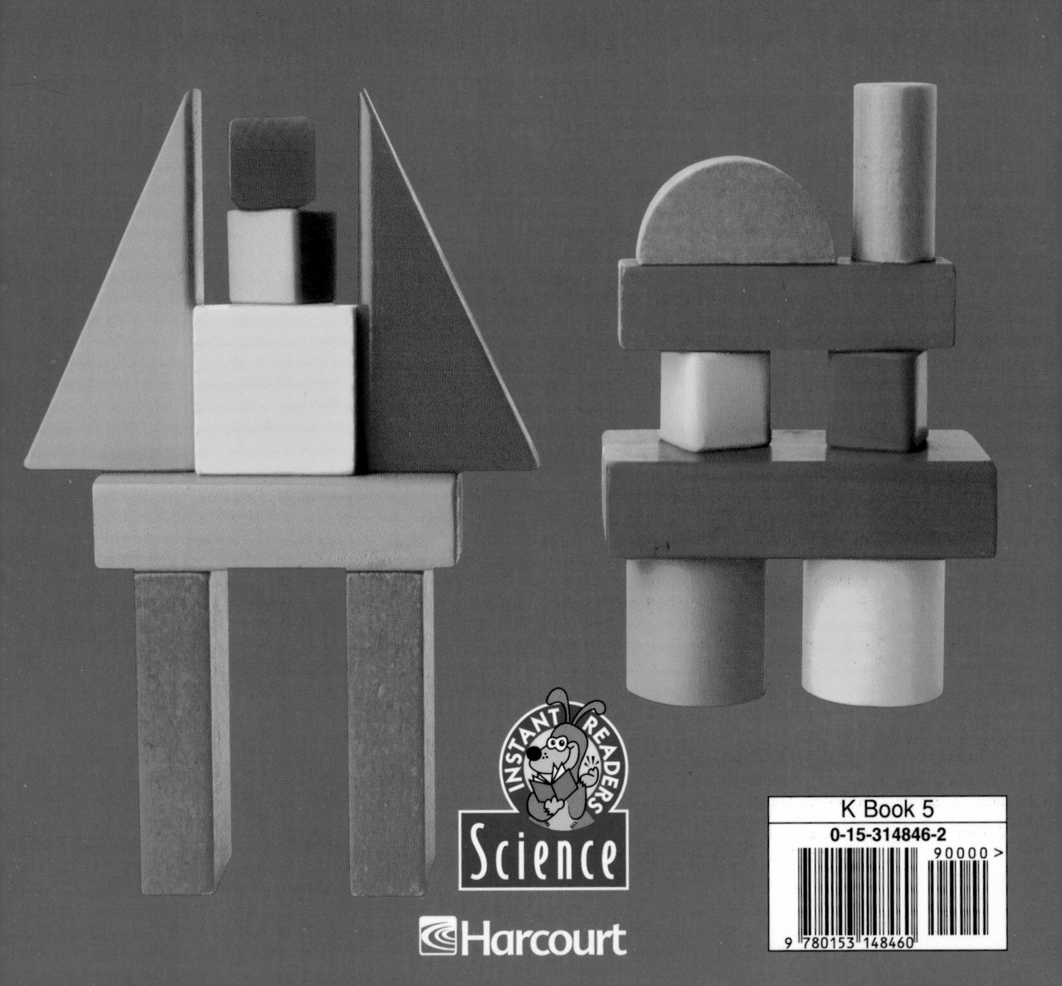

Harcourt

K Book 5
0-15-314846-2
90000 >
9 780153 148460

See the Seasons

Written by
Rozanne Lanczak Williams

THIS BOOK IS THE PROPERTY OF:

STATE ____________
PROVINCE ____________
COUNTY ____________
PARISH ____________
SCHOOL DISTRICT ____________
OTHER ____________

Book No. ______
Enter information in spaces to the left as instructed.

ISSUED TO	*Year Used*	*CONDITION*	
		ISSUED	*RETURNED*

PUPILS to whom this textbook is issued must not write on any page or mark any part of it in any way, consumable textbooks excepted.

1. Teachers should see that the pupil's name is clearly written in ink in the spaces above in every book issued.
2. The following terms should be used in recording the condition of the book: New; Good; Fair; Poor; Bad.